LIFE SCIENCES CONTRIBUTIONS
ROYAL ONTARIO MUSEUM
NUMBER 105

IAN R. BALL
N. GOURBAULT

The Morphology, Karyology and Taxonomy of a New Freshwater Planarian of the Genus *Phagocata* from California (Platyhelminthes: Turbellaria)

Publication date: 25 September 1975
ISBN 0–88854–181–3
Suggested citation: Life Sci. Contr., R. Ont. Mus.

ROYAL ONTARIO MUSEUM

PUBLICATIONS IN LIFE SCIENCES

The Royal Ontario Museum publishes three series in the Life Sciences:

LIFE SCIENCES CONTRIBUTIONS, a numbered series of original scientific publications, including monographic works.

LIFE SCIENCES OCCASIONAL PAPERS, a numbered series of original scientific publications, primarily short and usually of taxonomic significance.

LIFE SCIENCES MISCELLANEOUS PUBLICATIONS, an unnumbered series of publications of varied subject matter and format.

All manuscripts considered for publication are subject to the scrutiny and editorial policies of the Life Sciences Editorial Board, and to review by persons other than Museum staff who are authorities in the particular field involved.

IAN R. BALL is Assistant Curator in the Department of Entomology and Invertebrate Zoology, Royal Ontario Museum, and Adjunct Professor in the Department of Biology, University of Waterloo, Waterloo, Ontario.

NICOLE GOURBAULT is a CNRS Research Scientist in the Laboratoire de Zoologie (Vers), Museum National d'Histoire Naturelle, Paris, France.

PRICE: $1.50

100 Queen's Park, Toronto, Canada

PRINTED AND BOUND AT THE UNIVERSITY OF TORONTO PRESS

Contents

The Morphology, Karyology and Taxonomy of a New Freshwater Planarian of the Genus *Phagocata* from California (Platyhelminthes: Turbellaria)

Abstract

A new species of *Phagocata* from intermittent streams of Santa Barbara, California, is described. Morphologically it shows close affinities with Asian species of the genus. The karyotype differs from that of European and North American forms; comparative data are not available for the Asian species. Like *P. velata* in eastern North America, the new species withstands dry periods by a process of fragmentation and encystment. The nomenclature and classification of the genus *Phagocata* are discussed in some detail.

Introduction

The freshwater planarians of western North America are not well known (Kenk, 1972). This is unfortunate because the fauna is important from both a faunistic and biogeographical viewpoint. It seems likely, for example, that all representatives of the Planariidae and the Dendrocoelidae in North America have been derived from eastern Asia, whereas the more primitive Dugesiidae have been derived entirely from the south (Ball, 1974). In several papers Kenk (1953, 1973a, 1973b) has considered the relationships between Asian and western North American freshwater planarians, and in the present paper we add to this information by describing a new species of *Phagocata* from California that shows certain affinities with some Asian species of the genus. Karyological data also are provided; this is a field of study to which little attention has been paid with respect to North American planarians.

Materials and Methods

Numerous specimens were collected at various times from 1972 to 1974 by Mr. M. H. Fawcett from several intermittent streams in the vicinity of Santa Barbara, California. The specimens were fixed with saturated mercuric chloride in 0.9% sodium chloride and stored in 80% ethanol. In spring 1974 one batch of worms was received alive from Mr. Fawcett so that notes, photographs, and drawings of live specimens could be made. Attempts to culture the species in the laboratory proved fruitless.

Of several hundred specimens examined only four were mature or showed traces of reproductive organs. These were sectioned sagittally at 8 μm intervals, and the sections were mounted on 75 × 50 mm glass slides. The ribbons were mounted vertically, beginning in the top left corner with the slide label to the right. The sections were stained in phosphotungstic acid haematoxylin or chlorazol black-E, and mounted in DePex.

The description is based primarily on the holotype; this was fully mature and yielded the best set of sections. Features from the paratypes are confirmatory of the description.

Karyological data could be derived only from the live immature specimens received in May 1974. Consequently only mitosis in regenerative blastemas could be studied. Recently developed blastemas were immersed in 0.3% colchicine (Merck) for 3 to 4 hours, transferred to 2% acetic acid for 5 minutes, stained in lactic-aceto-orcein for 30 minutes, and then squashed under a cover-slip. The diploid complement was determined by examination of almost 30 metaphase plates. For nine very clear metaphase plates the relative lengths (length of chromosome × 100/total length of haploid genome) and the centromeric indices (length of shorter arm × 100/total length of the chromosome) of each of the chromosomes were calculated.

Means and standard deviations were also calculated. Whereas most elements were easy to pair, the numerical correspondencies from one plate to another were not always perfect — hence the high values of some of the standard deviations.

Systematic Section

***Phagocata fawcetti* sp. nov.**

Type Material

Department of Entomology and Invertebrate Zoology, Royal Ontario Museum. Holotype: sagittal sections on three slides (ROM C122). Paratypes: three sets of sagittal sections (ROM C75); immature specimens in alcohol (ROM C75, C122).

Diagnosis

Slender, pigmented, two eyes, head rounded with slight auricular lobes; anterior ramus of gut extends between eyes; penis bulb with small anterior and large posterior seminal vesicles; penis papilla broadly conical with terminal opening of ejaculatory duct; follicular testes ventral throughout the body length; bursa sacciform, common oviduct broad and long.

Etymology

The specific epithet is chosen to thank Mr. M. H. Fawcett (University of California, Santa Barbara), who worked so diligently on our behalf in collecting specimens.

General Features (Fig. 1A)

Slender pigmented species up to 15 × 1 mm; mature specimens usually about 11 × 0.7 mm. Chains of two or three individuals may reach 20 mm in length. The dorsal surface ranges in colour from dirty white to dark grey, the latter being characteristic of healthy well-fed individuals. The head is acutely rounded, sometimes with a slight projection in the mid-line, and with a distinct narrowing at the level of the eyes. The body margins are smooth and usually parallel, tapering to a pointed posterior end. There are two eyes, set in unpigmented areas and situated close together behind the anterior margin. For a body length of 100 units the positions of some important features are as follows: eyes 3.7 to 4.3, pharyngeal root 40–45, pharyngeal pore 65–70, genital pore 75–80. The anterior ramus of the intestine extends forwards between the eyes, but there are no pre-ocellar diverticula. The pharynx is of the normal planariid type.

Reproductive System (Fig. 1B)

The two ovaries present no peculiarities and are situated ventrally 11–14 from the anterior end for a body length of 100. The testes begin at about 20 and extend almost to the tail. The smaller follicles are predominantly ventral, but larger mature ones take up almost the entire dorso-ventral space. The penis is well developed and nearly fills the male atrium. There is a muscular hemispherical bulb, and from this a broad conical papilla projects ventrally posterior towards the gonopore. The papilla is covered with a flat nucleate epithelium that is underlain by circular and longitudinal muscle fibres. The circular fibres

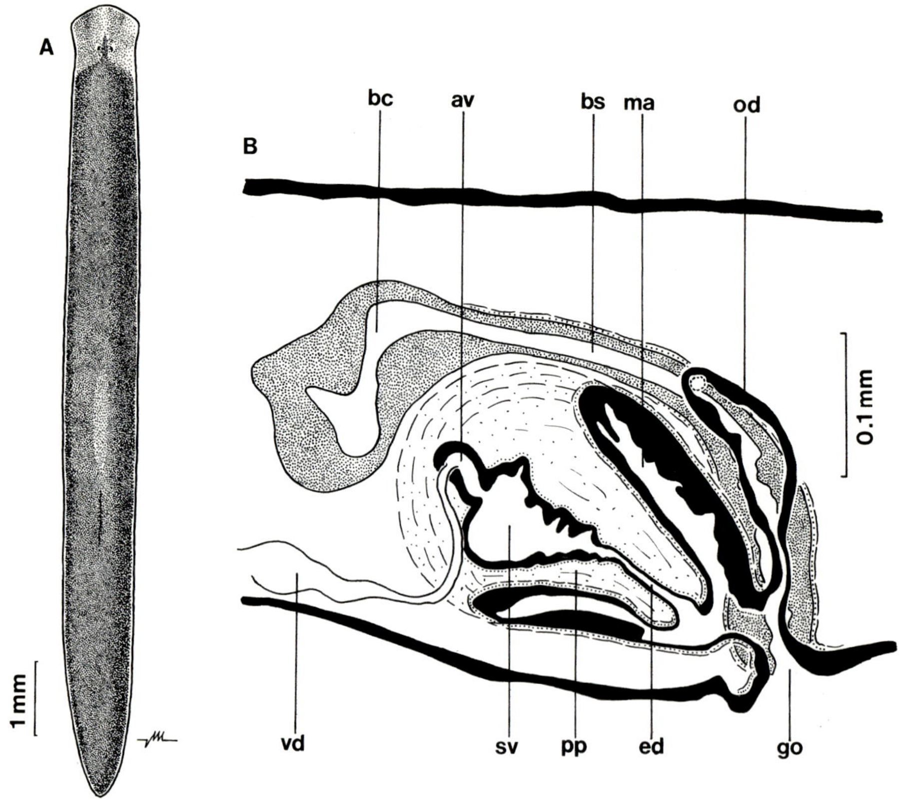

Fig. 1. *Phagocata fawcetti* sp. nov.

A. Dorsal view of living asexual specimen from the type locality.

B. Sagittal section of the copulatory apparatus of the holotype, viewed from the left side. Abbreviations: av, anterior seminal vesicle; bc, bursa copulatrix; bs, bursal canal; ed, ejaculatory duct; go, gonopore; ma, male atrium; od, common oviduct; pp, penis papilla; sv, posterior seminal vesicle; vd, left vas deferens.

are refractive and do not stain typically. They may be homologous with the unusual fibrous layer beneath the penial epithelium of *P. gracilis* (Kenk, 1970a). In the pharyngeal region the vasa deferentia enlarge to form the usual false seminal vesicles. At the level of the bursa they become thin once more, and at the penis bulb they turn dorsad and mediad, recurving slightly to enter the lateral walls of the bulb. Within the bulb they unite to form a small lobed vesicle lined with a low nucleate epithelium. This opens into a greatly expanded part of the ejaculatory duct that forms a posterior seminal vesicle. The ectal part of the ejaculatory duct is a narrow canal, displaced ventrally, and opening at the tip of the papilla. Both the posterior vesicle and the ejaculatory duct are lined with an irregular nucleate epithelium underlain with circular fibres. Numerous glands open into the posterior vesicle, and in the holotype its lumen is almost filled with granular secretion.

The bursa copulatrix is a sacciform organ lying anterior to the penis and adjacent to the wall of the penis bulb. The bursal canal passes over the copulatory apparatus as a narrow tube. It is lined with a ciliated, nucleate, cuboidal epithelium, which is overlain by a thin musculature of circular and longitudinal fibres. Posterior to the penis the bursal canal curves downwards, widens, and opens into the right side of the atrium at its junction with the gonopore. The cells lining the lumen of the ectal part of the bursal canal are much taller than elsewhere.

From the ovaries the oviducts run ventrally caudad to the copulatory apparatus. They curve around the walls of the male atrium and unite above the latter, without embracing the bursal canal, to form a common oviduct. The common oviduct is unusually long and broad and forms a somewhat elongate chamber lined with an irregular, nucleate, ciliated epithelium. It opens into the male atrium above the gonopore and somewhat to the left of the bursal canal. Numerous shell glands are associated with the ectal regions of the separate oviducts, though they appear to empty only into the common oviduct, principally into the upper portion.

In the available specimens the vitellaria are weakly developed, forming small follicles scattered throughout the mesenchyme.

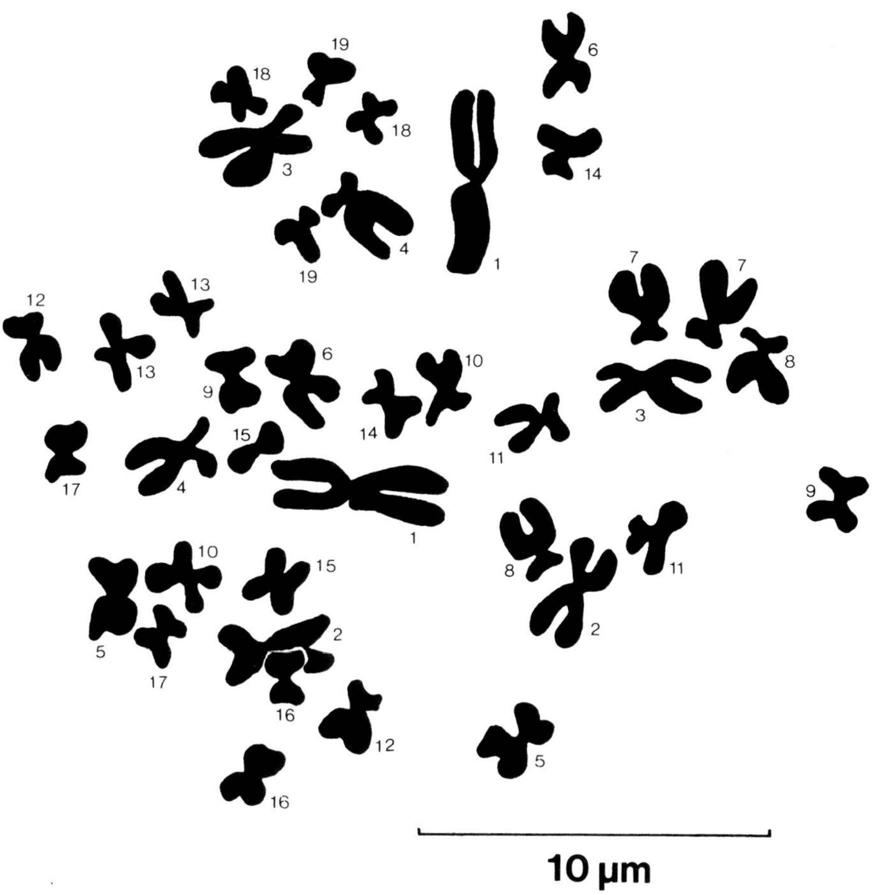

Fig. 2. *Phagocata fawcetti* sp. nov. Diagram of a metaphase plate.

Karyology (Figs. 2, 3)

The diploid complement consists of 38 elements (Fig. 2). The means of the relative lengths and the centromeric indices of these chromosomes, based on the nine clear metaphase plates, are given in Table 1 and represented in the idiogram forming Fig. 3. The karyotype exhibits one evident peculiarity, the presence of a pair of very large chromosomes which are more than 1.5 times the size of the second pair. In contrast, the remaining 17 pairs decrease in size very gradually, the last one being about half the size of the second pair.

With regard to centromere position most of the chromosomes are metacentric, though some are submetacentric (c.i. < 37)), according to the now classical nomenclature of Levan et al. (1964). The most metacentric chromosomes are the largest ones, and the pairs labelled 2, 3, and 18.

Table 1. Relative length and centromeric index of the 19 chromosome pairs of *Phagocata fawcetti*. Means and standard deviations from 9 metaphase plates.

Chromosome	Relative Length	Centromeric Index	Chromosome	Relative Length	Centromeric Index
1	11.37±0.58	49.21±0.92	11	4.69±0.14	35.24±5.10
2	7.54±0.37	43.06±1.93	12	4.54±0.21	40.90±7.35
3	6.41±0.23	43.48±2.38	13	4.40±0.19	39.55±4.77
4	5.95±0.24	35.79±6.15	14	4.22±0.17	36.98±4.33
5	5.70±0.08	42.13±5.58	15	4.12±0.12	37.83±3.60
6	5.42±0.15	37.28±4.44	16	4.05±0.12	35.33±6.41
7	5.30±0.18	34.03±6.66	17	3.91±0.18	41.39±5.14
8	5.02±0.17	41.48±3.62	18	3.72±0.26	45.85±2.30
9	4.88±0.18	42.87±5.02	19	3.54±0.14	36.71±1.76
10	4.82±0.18	40.36±5.97	—	—	—

Ecology

According to the collector (Fawcett, MS 1969) this species is abundant from November to April in streams near Santa Barbara that are not occupied by *Dugesia dorotocephala*. When active they glide very rapidly for planarians, and they seem primarily to be predators reacting quickly to the presence of damaged insects or earthworms. They rest well below the surface of the stream bottom, in gravel or plant roots. In April and May the worms encyst and remain encysted throughout the summer drought. In the live specimens studied in Toronto individuals often divided transversely, after losing much of their pigment, sometimes forming temporary chains of two or three pieces. This is precisely like the behaviour of *P. velata* before it encysts.

In eastern North America there are three planarians characteristic of temporary waters. *Hymanella retenuova* is a form closely related to *Phagocata*, and it produces a thick-shelled resting cocoon (Castle, 1941). *P. velata* and *P. vernalis* both encyst in the same way as the new species and in this way withstand dry periods. This phenomenon seems to be restricted to the genus *Phagocata* (Kenk, 1944).

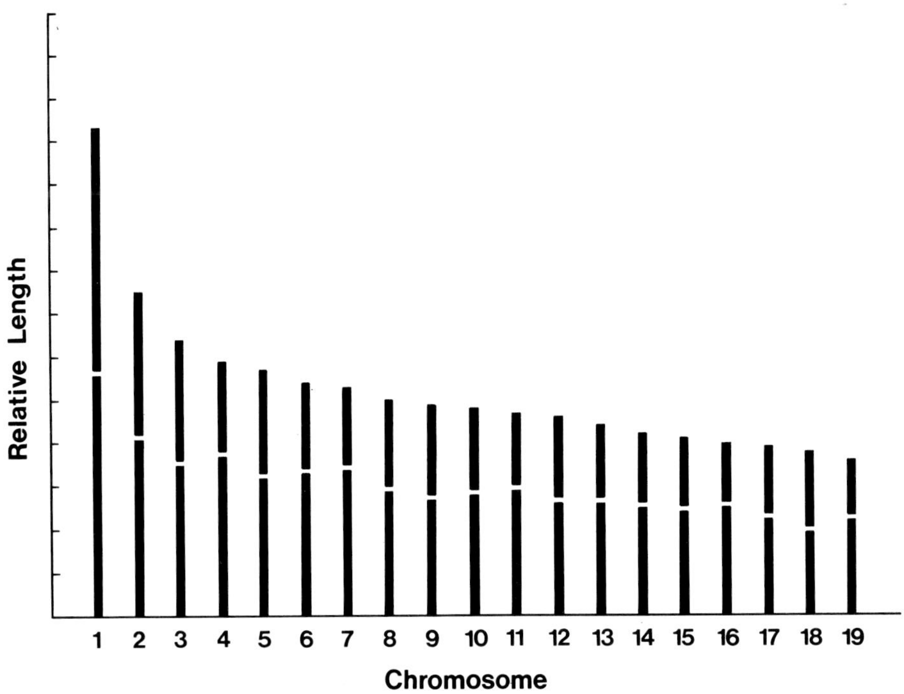

Fig. 3. *Phagocata fawcetti* sp. nov. Idiogram analysis based on the data in Table 1.

Distribution

California: Gaviota Creek, near Gaviota, Santa Barbara County (34°28′N, 120°14′W), October 1972 and subsequently, collector M. H. Fawcett, type locality; Quiota Creek, near Santa Ynez, Santa Barbara County (34°37′N, 120°05′W), 2 March 1973, collector M. H. Fawcett, many immature specimens, probably of this species.

Taxonomic Discussion

Morphological Affinities

As regards the North American fauna, *Phagocata fawcetti* needs to be compared with the five currently recognized pigmented monopharyngeal species of the genus (Kenk, 1972). These are *P. velata* (Stringer, 1909), *P. vernalis* Kenk, 1944, *P. tahoena* Kawakatsu, 1968, *P. crenophila* Carpenter, 1969, and *P. bulbosa* Kenk, 1970b. *P. monopharyngea* Hyman, 1945 is now known to be conspecific with *P. velata* (R. Kenk, pers. comm., 1974).

The principal differences between all these species reside in features of the reproductive apparatus, though there are differences apparent in external characters. This information may be summarized as follows:

P. fawcetti: Length up to 15 mm, length to breadth ratio 10–15:1. Head rounded, with slight lateral lobes. Testes predominantly ventral and throughout body length. Penis with well-developed bulb, large posterior seminal vesicle, strong broadly conical papilla, narrow ejaculatory duct opening terminally. Bursa sacciform. Common oviduct broad and long. California.

P. velata: Length up to 20 mm, length to breadth ratio 6–10:1. Testes predominantly dorsal. Ejaculatory duct with ventral diverticulum. Common oviduct of normal proportions. Eastern North America.

P. vernalis: Length up to 12 mm, length to breadth ratio 8:1. Testes fused, ventral, prepharyngeal. Penis bulb weak, without a well-developed vesicle. Bursa U-shaped. Common oviduct of normal proportions. Eastern North America.

P. tahoena: Length up to 12 mm, length to breadth ratio 6:1. Head rounded or truncate, without lateral lobes. Lumen of penis wide, not clearly differentiated into seminal vesicle and ejaculatory duct. Common oviduct of normal proportions. Lake Tahoe, Nevada/California.

P. crenophila: Length up to 22 mm, length to breadth ratio 8–10:1. Head variable. Penis bulb without well-developed vesicle, papilla plump with a dorsal protuberance. Ejaculatory duct opens ventrally. Common oviduct of normal proportions. Utah, Nevada, California.

P. bulbosa: Length up to 12 mm, length to breadth ratio about 7:1. Testes extend only to the mouth. Penis with slender conical papilla. Common oviduct of normal proportions. North Carolina.

Of all these species *P. fawcetti* is noteworthy for its extremely slender appearance. By this character alone the new species, when alive, can easily be distinguished from living *P. velata*, and according to Kenk (1970b) this latter species cannot be distinguished in life from *P. bulbosa, P. vernalis,* or *P. crenophila*. *P. tahoena* is an especially plump species (Kenk, 1970b: Fig. 7) and is quite different in appearance from *P. fawcetti*. However, judging from the description by Carpenter (1969), *P. crenophila*, known from cold springs and streams in northern California, Utah, and Nevada (Kenk, 1970b) is the species most similar to *P. fawcetti* in its long slender aspect. But the two species are very easily distinguished by the anatomy of the male copulatory organs.

In the anatomy of the male copulatory apparatus *P. fawcetti* shows some resemblances with *P. gracilis* (Haldeman, 1840). Thus both these species have an anterior and posterior seminal vesicle, the posterior one corresponding to the enlarged ental part of the ejaculatory duct, and both seem to have a peculiar fibrous layer beneath the penial epithelium. However, *P. gracilis* is polypharyngeal and is easily distinguished from *P. fawcetti*. It is also noteworthy that Kenk (1970b: 16) describes a fibrous layer, possibly of an elastic rather than muscular nature, beneath the penial epithelium of *P. bulbosa*. The differences between this species and *P. fawcetti* are outlined above.

Of the pigmented Asian species of the genus, *P. fawcetti* is similar to *P. iwamai* Ichikawa and Kawakatsu, 1962a, and *P. teshirogii* Ichikawa and Kawakatsu, 1962b, particularly in general body form and in the anatomy of the male reproductive organs. Superficially the Japanese species can be distinguished because in neither of them does the anterior ramus of the intestine extend between the eyes, as it does in the Californian species. More particularly, *P. teshirogii* differs from *P. fawcetti* in its more elongate and finger-like penis papilla and in the courses of the vasa deferentia and the form of the seminal vesicle; the common oviduct is of normal proportions. Phenetically *P. fawcetti* appears more closely related to *P. iwamai*, for in this latter species the ventral testes extend throughout the body length, and the penis papilla is a broad cone with an ejaculatory duct clearly differentiated into an ental cavity and an ectal canal, and it has a rather broad common oviduct (Ichikawa and Kawakatsu, 1962a: fig. 4). The principal differences between the species are that in *P. iwamai* the common oviduct is not as large as in *P. fawcetti* and is more anteriorly situated; in *P. iwamai* the vasa deferentia do not unite to form an anterior vesicle within the penis bulb; in *P. fawcetti* there is an unusual fibrous layer in the penial epithelium; and in *P. fawcetti* the anterior ramus of the gut extends between the eyes.

The principal peculiarity of the new species, the elongate and expanded common oviduct, is paralleled by two other Japanese species, *P. papillifera* Ijima and Kaburaki, 1916 and *P. suginoi* Kawakatsu, 1974 (in Kawakatsu et al., 1974). *P. papillifera*, as described by Kaburaki (1922), is unique amongst the three species in its possession of papillae on the dorsal surface of the body. It differs further from the new species in the form of the seminal vesicle which forms a simple bulbar cavity from which the narrow ejaculatory duct passes posteriorly, and, moreover, it is smaller, less slender, and much more lightly pigmented.

Phagocata suginoi is a small, delicate, and slightly pigmented species in which the anterior ramus of the intestine does not extend between the eyes (cf. Kawakatsu et al., 1974: fig. 1c), which latter are not set in pigment-free ocular areas. On all these counts *P. suginoi* may be distinguished from *P. fawcetti*. But in the anatomy and histology of the reproductive apparatus the two species are strikingly similar. The differences between them are subtle, and whereas Kawakatsu et al. (1974) have given a very detailed description of *P. suginoi*, it is evident from their published photomicrographs that their studies are based on contracted specimens. Principally, the differences between the species are as follows: in *P. suginoi* the penis is more slender and sinuous, the common oviduct is larger in that it equals or exceeds the penis in length, and the vasa deferentia enter into the primary seminal vesicle. It seems also that *P. suginoi* lacks the refractive fibrous elements in the wall of the penis papilla such as are found in *P. fawcetti*. Histologically the two species seem almost identical, with the exception that in *P. suginoi* the ectal thickening of the epithelium of the bursal canal is wanting. There can be little doubt that, of all extant *Phagocata* species, *P. suginoi* from Japan is the one most closely related to the new Californian species.

Karyological Affinities

There are no karyological data for Asian *Phagocata* species, and hitherto only two North American species of the genus have been analysed, by Whitehead (MS 1965), from this standpoint—*P. morgani* (Stevens and Boring, 1906) and *P. velata*. *P. morgani* seemed to have two biotypes, one diploid and one possibly polyploid. Sexually mature specimens had a diploid number of 14 chromosomes, with 7 bivalents indicated in meiosis. The polyploid specimens, with approximately 35 chromosomes, were obtained from populations where fission, without signs of sexual maturity, was observed. For *P. velata* Whitehead (op. cit.) recorded an approximate mitotic chromosome number of 60 elements, which might indicate polyploidy, and noted irregular meiosis. She also suggested that this was a recently evolved species and that pseudogamy may be involved in the reproductive cycle.

More exhaustive investigations of chromosome numbers and morphology have been carried out on four different European species of the genus *Phagocata* by Dahm (1958, 1964), and recently on some hypogeous populations of *P. vitta* (Dugès, 1830) from the south of France by Benazzi and Gourbault (1974). With respect to the existence of polyploidy, as well as to the occurrence of irregular meiosis, the data reported by Whitehead (MS 1965) seem generally consistent with the findings of Dahm (1958, 1964) for septentrional *P. vitta*. However, the karyology of the new species, *P. fawcetti*, certainly does not exhibit a polyploid condition, despite the rather high number of elements, because we can recognize just two chromosomes clearly differentiated in their length from all the others. This peculiarity has also been observed in most of the European species studied, *P. albissima* (Vejdovský, 1883), *P. ochridana* (Stanković and Komárek, 1927), and *P. vitta* from the south of France, though all these differ from *P. fawcetti* in their diploid complements of 36, 32, and 34 chromosomes respectively.

Whereas *P. fawcetti* might be expected to be taxonomically allied with *P. velata*, the two species differ greatly in their karyotypes. In fact *P. velata* possesses a varied and very high chromosome number of 60–96 elements corresponding to different degrees of polyploidy. In contrast the karyotype of *P. fawcetti* is quite similar to that of the type species of the genus, *P. gracilis*, although the two species are morphologically distinct. A detailed analysis of the karyotypes of *P. velata* and *P. gracilis* is in progress and will be published elsewhere.

The karyological data indicate once more that in the genus *Phagocata* specific differentiation seems to be linked with chromosomal variations, and we emphasize the opportunities that are made available to current planarian taxonomists by taking into consideration both morphological and karyotypical data. The evolutionary and biogeographic significances of our own data for *P. fawcetti* cannot be explored in further detail because we need to know more concerning the North American and, more especially, the Asian species of the genus.

The Status of the Genus *Phagocata*

The status of the genus *Phagocata* Leidy, 1847 has been the cause of much discussion among planarian specialists. The name was put forward as a subgenus of *Planaria* by Leidy (1847) for a polypharyngeal planarian, the *Planaria gracilis* of Haldeman (1840). The name, however, was little used and soon became largely forgotten. Komárek (1926), in his revision of the European freshwater planarians, divided the genus *Planaria* into several genera, including *Fonticola* and *Albiplanaria* for the close relatives of *Planaria vitta* and *Planaria albissima* respectively. Kenk (1930) combined these two genera, *Fonticola* being the valid name, and added to this unified genus several North American species that had a morphology similar to the European forms. He did not include *Phagocata gracilis* because according to the description by Woodworth (1897), which now is known to be erroneous (Hyman, 1937), the species belonged in the genus *Euplanaria* (=*Dugesia*). Later Hyman (1937) resurrected the name *Phagocata* and, finding no fundamental differences in the organization of the copulatory apparatus of *Phagocata gracilis* or *Phagocata woodworthi*, when compared with the European species of *Fonticola*, she concluded that they were one and the same genus with *Phagocata* as the prior and valid name.

Dahm (1949, 1958) accepted Hyman's decision and introduced the name *Phagocata* into the European literature, but some European workers disagreed and retained the name *Fonticola* for the European forms, either as a full genus (de Beauchamp, 1939, 1961; Reisinger, 1960) or as a subgenus of *Phagocata* (Gourbault, 1972). Many species have also been described from Asia, and these are placed under the name *Phagocata* (Kawakatsu, 1969). Dahm himself pointed out that the validity of Hyman's conclusion has never been properly assessed (Dahm, 1964), and Ball (1969) proposed that the priority principle invoked by Hyman was of doubtful validity and utility and that the name *Fonticola* should prevail for the genus in its broad sense. But this latter suggestion was not well received (Kawakatsu, 1969; Kenk, 1970a).

In the present paper we have adopted the conservative attitude of using the name *Phagocata*, but with recent advances in the theory of systematics the question now is not one of the priority of *Phagocata* over *Fonticola*, or vice versa, but concerns the very status of the genus as a whole. This question can be looked at in two ways, first by examining briefly the philosophy underlying definitions of taxa, and second by considering the validity of particular taxonomic characters.

If taxa are to be monophyletic, they must be defined by combinations of derived characters; primitive character states have little to say concerning genealogical relationships. Although this viewpoint has been acknowledged many times in the past, it was only recently that it was made explicit by the writings of Hennig (1966) and Brundin (1966). Both these authors have demonstrated convincingly that groups defined only by primitive characters are likely to be paraphyletic or polyphyletic.

The most recent definition of the genus *Phagocata* is that of Kenk (1970b:15): "Planariidae whose oviducts—without embracing the stalk of the bursa copulatrix—unite in a common oviduct which opens into the genital atrium. Male atrium without radial muscle plates. Adenodactyls absent. Eyes usually two." This definition holds even though the family Planariidae has been restricted by the exclusion of the Kenkiidae and of *Dugesia* and its allies (Ball, 1974), but it is deficient in that it fails to exclude *Hymanella* and *Plagnolia*. Kenk's definition essentially is polythetic because it includes no characters unique to the genus. In effect, *Phagocata* comprises all Planariidae that are not multiocular (cf. *Polycelis*), that do not have atrial muscle plates (cf. *Crenobia*), and that lack an adenodactyl (cf. *Planaria*); that is to say, until these latter genera had evolved the genus *Phagocata* did not exist. With respect to the Planariidae and Dendrocoelidae all the defining characters of *Phagocata* are in the primitive condition (Ball, 1974), and it is almost certain that it is a paraphyletic or polyphyletic group. To many this will be acceptable on the grounds of convenience, but it is not acceptable to those concerned with elucidating the evolutionary and biogeographical relationships of these animals. Yet Kenk (1935, 1953) has recognized that the genus is not homogeneous and probably will be divided; this also is apparent in the writings of European workers (de Beauchamp, 1939; Reisinger, 1960; Dahm, 1958, 1964; Gourbault, 1969, 1972). There seems little doubt that in many respects the type species of *Phagocata*, *P. gracilis*, is quite different from the type species of *Fonticola*, *F. olivacea*, and the retention of *Fonticola* for the European forms seems justified.

The recognition of the philosophical and methodological importance of distinguishing between derived and ancestral character states makes possible a reassessment of some views concerning taxonomic characters within *Phagocata* and related genera. Originally *Phagocata* was defined on the basis of polypharyngy and if this criterion were still applied a group comprising two very distinct species, *P. gracilis* and *P. woodworthi*, would be formed. However, the taxonomic significance of polypharyngy has been questioned by most subsequent workers, the most recent discussion being that of Kenk (1970a). The latter author argues that it cannot be a character of generic significance in *Phagocata* because it is not so considered in the genera *Crenobia* and *Sphalloplana*, both of which have polypharyngeal members. Indeed, Kenk points out that the polypharyngeal *Crenobia montenigrina* and *C. teratophila* are often considered to be but subspecies of the monopharyngeal *C. alpina*. Moreover, the polypharyngeal forms of *Sphalloplana* agree in general anatomical features with the monopharyngeal members of the genus. This is true, but it is misleading to make direct comparison between these genera and *Phagocata* because the polypharyngeal *Crenobia* species and the polypharyngeal *Sphalloplana* species are readily assigned to their respective genera by suites of derived character states—character states unique to those genera. This is not the case with the comparable *Phagocata* species. Nor is it valid to argue, though taxonomically it may be convenient to do so, that what is unacceptable in one genus also is unacceptable in another genus. If this were the case, then havoc would be created in the planarian system, for adenodactyls are considered to be unimportant in the supraspecific classification of *Dugesia* and *Polycelis* species, yet are the defining character of *Planaria*, a genus closely related to *Phagocata* s.l. And this is but one example of many that could be quoted. It follows that the logical validity of defining *Phagocata* on the basis of polypharyngy has not been effectively challenged.

But if *Phagocata* is to be restricted to the polypharyngeal forms, the problem remains of how to deal with the Asian and remaining North American species. To continue to lump them indiscriminately into *Phagocata* seems misleading, but a more rational approach requires a rethinking of taxonomic ideas and a search for new classes of data relevant to the problem. Cytogenetic work on Asian and American faunas, for example, seems indispensable, and we need to be more critical in our assessment of features found in individual species. The peculiar fibro-muscular tissue found in some forms is a case in point, for we do not know its taxonomic distribution, and we should look at characters other than those associated with the copulatory apparatus.

In assigning to our new species the generic name *Phagocata* we have taken the path of least resistance; it would be premature to introduce any new generic names. It is more than likely that the nomenclature of this, and related, species will be subject to emendation when the availability of adequate material makes possible a thorough revision of the genus in its broad sense.

Acknowledgments

It is a pleasure to thank Mr. M. H. Fawcett, University of California at Santa Barbara, for collecting specimens on our behalf and for donating them to us. We thank also Dr. Roman Kenk (USNM) for permission to quote some unpublished data, Mr. Julian Mulock (ROM) for drawing Figure 1A, Ms. Maria Tran Thi Vinh-Hao for drawing Figure 3, and Rev. J. C. E. Riotte for the German translation of the summary. Ms. Jennifer Smith typed several versions of the manuscript.

Dr. Gourbault's visit to the ROM was made possible by an award from the National Research Council of Canada under its Canada-France Cultural Exchange Programme. We acknowledge with gratitude this support.

Summary

A new pigmented planarian, *Phagocata fawcetti* sp. nov., is described from intermittent streams of southern California. Morphologically the species is most similar to the Japanese species *P. suginoi*, especially in habitus and in the anatomy of the reproductive organs. Of the North American *Phagocata*, species *P. fawcetti* is most similar to *P. crenophila* in habitus and to the polypharyngeal *P. gracilis* in the anatomy of the male copulatory organ, and in its karyotype. Criteria for distinguishing all these species are given.

The populations studied consist of diploid individuals with a diploid set of 38 chromosomes, most of which are metacentric though a few are submetacentric. Only two other North American *Phagocata* species have been studied karyologically, and these are very different from *P. fawcetti*. Different degrees of polyploidy occur in *P. velata*, which has a basic complement of 60–96 chromosomes; and in *P. morgani*, with a haploid complement of 7, there are both diploid and polyploid races. Comparative data for Asian species are lacking.

The karyotype of *P. fawcetti* includes two chromosomes clearly differentiated by length from all the others. This peculiarity has also been noted in the European species *P. albissima*, *P. ochridana*, and *P. vitta*, but these forms differ in their diploid complements of 36, 32, and 34 chromosomes respectively.

P. fawcetti is one of three species in North America that inhabit temporary waters. It withstands dry periods by a process of fragmentation and encystment, as does *P. velata*.

The generic name *Phagocata* is applied to the new species with some reservation because the genus is defined only by primitive characters and is unlikely to be a strict monophyletic group. However, until a major revision of the entire complex is undertaken, it would be unwise to introduce new generic names.

Sommaire

Phagocata fawcetti, espèce nouvelle de Triclade, pigmentée, provient de cours d'eau temporaires du Sud de la Californie. D'un point de vue morphologique, cette espèce est proche de la Planaire japonaise *P. suginoi*, à la fois par son habitus et l'anatomie de ses organes reproducteurs.

Parmi ses congénères nord-américaines, cette *Phagocata* s'apparente particulièrement à *P. crenophila* par son habitus et aussi à *P. gracilis*, espèce polypharyngée, par l'anatomie de l'appareil copulateur male et par son caryotype. Une liste des critères permettant de distinguer ces espèces est donnée.

Les individus des populations étudiées possèdent une garniture chromosomique diploide de 38 éléments métacentriques pour la plupart, un petit nombre étant submétacentriques.

Jusqu'ici seules deux autres espèces nord-américaines de *Phagocata* ont donné lieu à des études caryologiques; les résultats diffèrent de ceux obtenus pour *P. fawcetti*: *P. velata* est polyploide (60 à 96 chromosomes), et *P. morgani* (avec un nombre de base 7) comprend des races soit diploides, soit polyploides. On ne possède aucune donnée concernant les espèces asiatiques.

Le caryotype de *P. fawcetti* montre deux chromosomes nettement différents de tous les autres par leur longueur; cette particularité a déjà été notée pour les espèces européennes *P. albissima*, *P. ochridana*, et *P. vitta*, mais ces dernières possèdent des nombres diploides différents, respectivement 36, 32, et 34.

P. fawcetti est une des trois espèces nord-américaines peuplant des eaux temporaires. Elle résiste aux périodes de sécheresse en se fragmentant et en s'enkystant comme le fait *P. velata*.

Cette nouvelle espèce est attribuée au genre *Phagocata* avec certaines réserves, car ce genre n'est défini que par des caractères primitifs et ne constitue pas vraiment un groupe monophylétique. Toutefois, tant qu'une révision complète de l'ensemble des formes n'est pas effectuée, il semble inopportun de créer de nouveaux genres.

Zusammenfassung

Eine neue Art pigmentierter Planarien, *Phagocata fawcetti* sp. nov., ist beschrieben von zeitweilig austrocknenden Wasserläufen in Südkalifornien. In ihrer Morphologie ist die neue Art am meisten der japanischen Art *P. suginoi* ähnlich, besonders in Bau und Anatomie der männlichen Fortpflanzungsorgane. Unter den nordamerikanischne Arten von *Phagocata* ist die neue Art am meisten der *P. crenophila* in Bau und der polypharyngealen *P. gracilis* in der Anatomie der männlichen Kopulationsorgane sowie in ihrer Karyotype ähnlich. Kriterien für die Unterscheidung aller dieser Arten sind gegeben.

Die untersuchten Populationen bestehen aus diploiden Individuen mit einem diploiden Chromosomsatz von 38, von denen die meisten metazentrisch sind neben einigen wenigen submetazentrischen. Nur zwei andere nordamerikanische *Phagocata* Arten sind karyologisch untersucht worden und diese sind sehr verschieden von *P. fawcetti*. Verschiedene Grade von Polyploidie kommen in *P. velata* vor, die ein Komplement von 60–96 Chromsomen hat; in *P. morgani*, mit einem haploiden Komplement von 7, gibt es beides, diploide und polyploide Rassen. Vergleichbare Daten fur asiatische Arten liegen nicht vor.

Die Karyotype von *P. fawcetti* schliesst zwei Chromosome mit ein, die durch ihre Länge klar unterschieden sind von all den anderen. Diese Besonderheit ist auch in den europäischen Arten *P. albissima*, *P. ochridana* und *P. vitta* festgestellt worden. Diese Formen sind jedoch verschieden in ihren jeweiligen diploiden Komplementen von 36, 32, und 34 Chromosomen.

P. fawcetti ist eine von drei nordamerikanischen Arten, die temporäre Wasserläufe bewohnen. Sie überdauert Trockenperioden durch einen Fragmentations- und Einkapselungsprozess, genau so wei in *P. velata*.

Der Gattungsname *Phagocata* ist für die neue Art mit einiger Reservation gebraucht, da die Gattung nur durch primitive Charaktere umschrieben ist und kaum eine streng monophyletische Gruppe bildet. Es würde jedoch unangebracht sein, einen neuen Gattungsnamen einzuführen, bevor eine grössere Revision des Gesamtkomplexes vorgenommen worden ist.

Literature Cited

BALL, I. R.

1969 An annotated checklist of the freshwater Tricladida of the Nearctic and Neotropical regions. – Canadian Journal of Zoology, 47: 59–64.

1974 A contribution to the phylogeny and biogeography of the freshwater triclads (Platyhelminthes, Turbellaria). – *In* Riser, N. W. and M. P. Morse, Eds. Biology of the Turbellaria. New York, McGraw-Hill, pp. 339–401.

BEAUCHAMP, P. DE

1939 La systématique et l'éthologie des *Fonticola* (Turbellaria, Triclades). – Věstník Československé Zoologické Společnosti v Praze, 6–7: 91–96.

1961 Classe des Turbellariés. – *In* Grassé P.P. éd. Traité de Zoologie, Vol. 4. Paris, Masson, pp. 35–212.

BENAZZI, M. and N. GOURBAULT

1974 Étude caryologique de quelques populations hypogées de la planaire *Phagocata* (*Fonticola*) *vitta* (Dugès, 1830.) – Caryologia, 27: 467–484.

BRUNDIN, L.

1966 Transantarctic relationships and their significance, as evidenced by chironomid midges, with a monograph of the subfamilies Podonominae and Aphroteniinae and the austral Heptagyiae. – Kungliga Svenska Vetenskapsakademiens Handlingar, Ser. 4, 11: 1–472.

CARPENTER, J. H.

1969 A new planarian from Utah, *Phagocata crenophila* n. sp. (Turbellaria, Tricladida). – Transactions of the American Microscopical Society, 88: 274–281.

CASTLE, W. A.

1941 The morphology and life history of *Hymanella retenuova*, a new species of triclad from New England. – American Midland Naturalist, 26: 85–97.

DAHM, A. G.

1949 Phagocata (=*Fonticola*) from South Sweden (Turbellaria Tricladida Paludicola). – Lunds Universitets Årsskrift, N.F. Avd. 2. 45(7): 1–32.

1958 Taxonomy and Ecology of Five Species Groups in the Family Planariidae (Turbellaria, Tricladida, Paludicola). – Malmö, 241 pp.

1964 The taxonomic relationships of the European species of *Phagocata* (?=*Fonticola*) based on karyological evidence. – Arkiv för Zoologi, ser 2, 16(24): 481–509.

DUGÈS, A.

1830 Aperçu de quelques observations nouvelles sur les Planaires et plusieurs genres voisins. – Annales des Sciences Naturelles, 21: 72–92.

FAWCETT, M. H.

MS1969 Factors influencing the distribution of two stream-dwelling triclads near Santa Barbara, California. – M.A. Thesis, University of California, Santa Barbara.

GOURBAULT, N.

1969 Triclades obscuricoles des Pyrénées. III. Contribution à l'étude des *Atrioplanaria*: *Atrioplanaria delamarei* sp. n. – Annales de Spéléologie, 24: 377–385.

1972 Recherches sur les triclades paludicoles hypogés. – Mémoires du Muséum national d'Histoire Naturelle, Paris, Sér. A. 73: 1–249.

HALDEMAN, S.

1840 Supplement to Number One of "A monograph of the Limniades or fresh-water univalve shells of North America". – Philadelphia, 3 pp.

HENNIG, W.

1966 Phylogenetic Systematics. – Urbana, University of Illinois Press, 263 pp.

HYMAN, L. H.
1937 Studies on the morphology, taxonomy, and distribution of North American triclad Turbellaria. VII. The two species confused under the name *Phagocata gracilis*, the validity of the generic name *Phagocata* Leidy 1847, and its priority over *Fonticola* Komárek 1926. – Transactions of the American Microscopical Society, 56: 298–310.
1945 North American triclad Turbellaria. XI. New, chiefly cavernicolous planarians. – American Midland Naturalist, 34: 475–484.

ICHIKAWA, A. and M. KAWAKATSU
1962a *Phagocata iwamai*, a new freshwater planarian, from Hokkaido. – Annotationes Zoologicae Japonenses, 35: 38–46.
1962b *Phagocata teshirogii*, a new freshwater planarian, from the Tôhoku region in Honshû. – Annotationes Zoologicae Japonenses, 35: 112–120.

IJIMA, I. and T. KABURAKI
1916 Preliminary descriptions of some Japanese triclads. – Annotationes Zoologicae Japonenses, 9: 153–171.

KABURAKI, T.
1922 On some Japanese freshwater triclads; with a note on the parallelism in their distribution in Europe and Japan. – Journal of the College of Science, Tokyo Imperial University, 44(Art.2): 1–72.

KAWAKATSU, M.
1968 North American triclad Turbellaria. XVIII. Freshwater planarians from Lake Tahoe. – Proceedings of the United States National Museum, 124: 1–21.
1969 An illustrated list of Japanese freshwater planarians in color. – Bulletin of the Fuji Women's College, 7 (ser. II): 45–91.

KAWAKATSU, M., H. MURAYAMA and F. NÎMURA
1974 *Phagocata suginoi* sp. nov., a new probably subterranean freshwater planarian from the Hokuriku District in Honshû. – Annotationes Zoologicae Japonenses, 47: 147–159.

KENK, R.
1930 Beiträge zum System der Probursalier (Tricladida paludicola). – Zoologischer Anzeiger, 89(5/6): 145–162; 89(11/12): 289–302.
1935 Studies on Virginian triclads. – Journal of the Elisha Mitchell Scientific Society, 51(1): 79–133.
1944 The fresh-water triclads of Michigan. – Miscellaneous Publications of the Museum of Zoology, University of Michigan, 60:1–44.
1953 The fresh-water triclads (Turbellaria) of Alaska. – Proceedings of the United States National Museum, 103: 163–186.
1970a Freshwater triclads (Turbellaria) of North America IV. The polypharyngeal species of *Phagocata*. – Smithsonian Contributions to Zoology, 80: 1–17.
1970b Freshwater triclads (Turbellaria) of North America. II. New or little known species of *Phagocata*. – Proceedings of the Biological Society of Washington, 83: 13–34.
1972 Freshwater planarians (Turbellaria) of North America. – Biota of Freshwater Ecosystems, Identification Manual No. 1. Washington, D.C., Environmental Protection Agency. 81 pp.
1973a Freshwater triclads (Turbellaria) of North America, V: The genus *Polycelis*. – Smithsonian Contributions to Zoology, 135: 1–15.
1973b Freshwater Triclads (Turbellaria) of North America, VI: The genus *Dendrocoelopsis*. – Smithsonian Contributions to Zoology, 138: 1–16.

KOMÁREK, J.
1926 Dolpněk ku Vejdovského revisi českých Triclad. – Věstník Královské České Společnosti Nauk, Třída matematicko-přírodovědecká, Sbornik Vejdovského, 1925(7): 1–32.

LEIDY, J.
1847 Description and anatomy of a new and curious sub-genus of *Planaria*. – Proceedings of the Academy of Natural Sciences of Philadelphia, 3: 248–251.

LEVAN, A., K. FREDGA and A. A. SANDBERG
1964 Nomenclature for centromeric position on chromosomes. – Hereditas, 52: 201–220.

REISINGER, E.
1960 Vitale Nervenfärbungen bei Plathelminthen und ihre Abhängigkeit vom physiologischen Zustand des Organismus. – Zeitschrift für wissenschaftliche Zoologie, Abteilung A, 164: 271–293.

STANKOVIC, S. and J. KOMÁREK
1927 Die Süsswasser-Tricladen des Westbalkans und die zoogeographischen Problem dieser Gegend. – Zoologische Jahrubücher, Abteilung für Systematik Ökologie und Geographie der Tiere, 53: 591–674.

STEVENS, N. M. and A. BORING
1906 *Planaria morgani* n. sp. – Proceedings of the Academy of Natural Sciences of Philadelphia, 58: 7–9.

STRINGER, C. E.
1909 Notes on Nebraska Turbellaria with descriptions of two new species. – Zoologischer Anzeiger, 34: 257–262.

VEJDOVSKY, F.
1883 Exkrécní apparát Planarií. – Zprávy o Zasedání České Společnosti Nauk v Praze, 1882: 273–280.

WHITEHEAD, M. M.
MS 1965 The triclads of Cattaraugus County, New York. – Ph.D. Thesis, Saint Bonaventure University, N.Y. (University Microfilms Inc.). 128 pp.

WOODWORTH, W. McM.
1897 Contributions to the morphology of the Turbellaria. II. On some Turbellaria from Illinois. – Bulletin of the Museum of Comparative Zoology at Harvard College, 31: 1–16.